Winfield Shoop

The Ready Reckoner, Calculator, and Mechanic's Companion

For lumber dealers, carpenters, mechanics, masons, farmers, merchants,

laborers, &c.

Winfield Shoop

The Ready Reckoner, Calculator, and Mechanic's Companion
For lumber dealers, carpenters, mechanics, masons, farmers, merchants, laborers,
&c.

ISBN/EAN: 9783337370589

Printed in Europe, USA, Canada, Australia, Japan

Cover: Foto ©berggeist007 / pixelio.de

More available books at **www.hansebooks.com**

THE
READY RECKONER,
CALCULATOR,
AND
Mechanic's Companion.

FOR
LUMBER DEALERS, CARPENTERS, MECHANICS, MASONS, FARMERS, MERCHANTS, LABORERS, &C.

ALSO

Including Form for
MECHANICS' LIEN

BY

· WINFIELD SHOOP.

CHAMBERSBURG, PA.:
AT THE OFFICE OF THE "VALLEY SPIRIT,"
1875.

Recommendations.

We have carefully examined the Ready Reckoner, Calculator and Mechanic's Companion, by Winfield Shoop, and we take pleasure in recommending it to the public and all those that are interested in such books, as being a correct and useful work.

P. A. WITMER, Examiner of Schools for Washington County, Md.

D. O. WITMER, Principal Williamsport Grammar School.

THOMAS E. WILLIARD, Register of Wills for Washington County, Md.

GRUBER & WITMER, Proprietors Steam Saw and Planing Mill, Williamsport, Md.

STEFFEY & IRWIN, Lumber Merchants, Williamsport, Md.

J. L. MOTTER, Teller Washington County National Bank, Williamsport, Md.

GEO. P. LEITER, Clerk Washington County Commissioners.

Contents

Scantling Measure.

RULE.

Multiply the depth by the breadth and the product by the length, and divide by 12.

EXAMPLE.

A piece of timber 4 x 5 inches, and 18 feet long, how many feet does it contain?

5 inches.	18 feet long.
4 "	20
20	12)360

Ans. 30 feet in piece.

Scantling, Rafters &c.

RULE.

Multiply both ends, add together, take half the product, multiply by the length and divide by 12.

EXAMPLE.

A piece of timber 4 x 5 inches at one end and 3 x 6 at the other end and 12 feet long.

4 inches.	6 inches.
5 "	3 "
20 "	18 "
	20

Take half 38

19 inches average.
12 feet long.

12)228

Ans. 19 feet in piece.

Board and Plank Measure.

RULE.

Multiply the width of board or plank by the length of same and divide by 12. If the board is more than an inch thick add accordingly.

EXAMPLE No. 1.

A board 12 inches wide, 16 feet long, and 1 inch thick.

```
12 inches wide.
16 feet long.       12)192
 —                  ——
 72                 Ans.  16 feet in board.
 12
 ——
192
```

EXAMPLE No. 2

A board 18 inches wide, 16 feet long and 1¼ inches thick.

```
  18 inches wide.
  16 feet long.      12)288
  —                  ——
108                     24
 18                Add ¼  6
 ——                     —
288                Ans. 30 feet.
```

EXAMPLE No. 3.

A plank 12 inches wide, 16 feet long and 1¾ inches thick.

```
 16 feet long.      12)192
 12 inches.         ——
 —                     16
192               Add ¾ 12
                       ——
          Ans.        28 feet.
```

Measuring Boards and Plank with a Tape.

RULE.

In measuring lumber with a tape line, if the
boards are 16 feet long and 1 inch thick and you
have 46 feet 9 inches line then multiply the
length of the line by the length of the boards,
or in other words, always multiply the length
of the line by the length of the boards you are
measuring, and if the boards are more than an
inch in thickness add according as in preceeding
page.

EXAMPLE No. 1.

```
46 feet 9 inches line      16 feet long.
16 feet long.                 9 inches
___                          ___
276                        12)144
46                            ___
 12 feet in 9 inches.         12 feet

748     Ans. 748 feet boards.
```

EXAMPLE No. 2.

Forty feet of line. Boards 18 feet long and 1¼
inches thick.

```
18 feet long.               720
   40 feet line.   Add ¼    180
___                        ___
720                  Ans.  900 feet.
```

Square Timber.

RULE.

Multiply the breadth by the breadth and the product by the length and divide by 12

EXAMPLE.

A piece of timber 12 x 12 inches and 16 feet long.

```
 12            144
 12             16
---            ---
144            864
               144
              ----
           12)2304
               ----
Ans.           192 feet.
```

Round Timber.

RULE.

Square the mean diameter and multiply the product by 7854; Cut off four right hand figures; Multiply by the length and divide by 12.

EXAMPLE.

A log 14 inches mean diameter and 20 feet long.

```
14 inches.         7854
14    "             196
---              ------
 56             47124
 14             70686
---              7854
196            -------
             153|9384
              20 feet long.
             -----
          12)3060
             ------
Ans.          252 feet in log.
```

To Find the Mean Diameter of Logs.

RULE.

To ascertain the mean diameter of logs, measure both ends inside the bark; add together; take half the product to get the average or mean diameter of log.

EXAMPLE.

A log 44 inches in diameter at one end and 36 inches in diameter at the other end. What is the mean diameter.

```
44 inches at one end.
36   "    at other end.
      —
Half)80
      —
```

Ans.   40 inches in mean diameter.

To Square Round Timber.

RULE.

Multiply the diameter by 70 and cut off two right hand figures.

EXAMPLE.

A log 20 inches diameter. What will it square.

```
20 inches.
70
——
```

Ans.   14|00   Square 14 inches.

Cubic Feet in Round Timber.

RULE.

Square the mean diameter; Multiply the product by 7854; Cut off four right hand figures; Multiply by the length and divide by 144.

EXAMPLE.

A log 18 inches mean diameter and 34 feet long.

```
  18                          7854
  18                            34
 ───                         ─────
 144                         31416
  18                         15708
 ───                         23562
 324
                          254|4896
                             34 feet long.

                            1016
                             762
                            ────           Ans.
                      144)8630(59 Cubic feet and
                          720        140 inches.

                           1436
                           1296

                            140 inches.
```

Cubic Feet in Square Timber.

RULE.

Multiply the breadth by the depth in inches, and the product by the length, and divide by 144.

EXAMPLE.

A piece of timber 15 x 16 inches square and 40 feet long. How many Cubic feet.

```
15 inches broad.
16  "    deep.                240
 —                            40 feet long.
90                            —
15                   144)9600(66  Cubic ft.
 —                        864       & 96 in.
240                       —
                          960
                          864
                          —
                          96 Inches.
```

Ans. 66 Cubic feet and 96 inches.

To Tell the Height of a Tree while Standing.

RULE.

Multiply the length of the shadow of a tree by the height of a stick and divide by the shadow of the stick.

EXAMPLE.

Suppose the shadow of a tree 90 feet ; Stick 4 feet long, and the shadow of stick 6 feet long. What is the height of tree.

```
                90 feet shadow of tree.
   Shadow         4  "   height of stick.
   of stick      —
       6 feet)360

   Ans.         60 feet tree's height.
```

Wood and Bark Measure.

RULE.

Multiply the length, breadth and depth together and divide the product by 128.

EXAMPLE.

A pile of wood or bark 150 feet long, 8 feet high and 4 feet wide. How many cords.

$$
\begin{array}{r}
150 \text{ feet long.} \\
8 \text{ `` high.} \\
\hline
1200 \\
4 \text{ feet wide.} \\
\hline
\end{array}
$$

$$128)4800(37\tfrac{1}{2} \text{ cords.}$$
$$\underline{384}$$
$$960$$
$$\underline{896}$$
$$64$$
$$\frac{\quad}{128} = \tfrac{1}{2}$$

Ans. 37½ Cords.

Bark and Wood Measure—Irregular Size.

RULE.

Add the breadth and height together. Take half the sum to get the average width, then multiply the average width by height and length and divide by 128 as in preceeding rule.

EXAMPLE.

A pile of wood or bark 8 feet wide at top, 4 feet wide at bottom, 6 feet high and 9 feet long. How many cords.

```
              8 feet top.
              4  "   bottom.
              --
half         12
              --
              6 feet average.
              6  "   high.
              --
             36
              9 feet long.
              --
        128)324(2¼ cords Ans.
            256
             --
             68
            ---=¼
            128
```

Bushels Corn Crib Contains.

RULE.

Multiply the length by the breadth and the product by the depth, then multiply that product by 4 and divide by 5.

EXAMPLE.

A crib 30 feet long, 6 feet broad and 10 feet high. How many bushels shelled corn does it contain.

```
30 feet long.          1800
 6  "  broad.             4
---                    ------
180                    5)7200
   10 feet high.       ------
---                    1440  bus. Ans.
1800
```

To get the number flour barrels, divide the number bushels by three, and the number of regular corn barrels divide by 5.

Number of Brick Pavement Contains.

RULE.

Multiply the length by the breadth and that product by 4½.

EXAMPLE.

A pavement 40 feet long and 9 feet wide.

```
40 feet long.            360
 9  "   wide              4½
 —                        —
360 square feet.         1440
                          180
                          —
              Ans.       1620  Brick.
```

```
9)360 square feet.
  —
  40    "    yards.
  40½ brick to square yard.
  —
  1600
    20
  —
  1620 brick.
```

Stone Masonry.

RULE.

Multiply the length by the height and that product by the width and divide by 24¾.

EXAMPLE.

A wall 42 feet 3 inches long, 8 feet high and 2 feet thick. How many perches?

```
42 ft. 3 in. long.              3 inches. 
 8 ft. high.                    8 feet
 ──                             ──
336                          12)24
 2 feet in 3 inches.            ──
 ──                              2 feet.
338                     338
                          2 feet thick.
                        ──
              24¾  676
               4     4
               ──    ──
              99  )2704(27⅓
                   198
                   ───
                   724
                   693
                   ───
                    31
                    ──=⅓
                    99
```

Ans. 27⅓ Perches.

2.

Brick Masonry.

RULE.

Bring the whole into feet, as in preceding rule and if the wall is 9 inches thick (in order to get the number of brick house contains) multiply the aggregate number of feet the house contains by 12; if the wall is 14 inches thick, multiply the same by 18, and if 18 inches thick, multiply by 24, and for every 4 inches of wall add 6 to the multiplier.

EXAMPLE.

A building 25 feet long, 20 feet broad and 15 feet high, and gable end 12 feet from centre of wall to comb of roof, the wall of same 14 inches thick; how many brick does the house contain?

```
 25 feet long.
 15  "  high.
 ───
 125
  25
 ───
 375 feet one side.
   2
 ───
 750 feet two sides.
 ───
  15 feet high.
  20  "  broad.
 ───
 300  "  one side.
   2
 ───
 600 feet  two sides.
 750
 ───
1350 feet in four sides.
```

```
       20 feet broad
       12
       ───
       240 ft. both ends.
      1350 "  four sides.
       ───
      1590   ft. house con-
        18   tains.
       ───
      12720
      1590
       ───
      28620  Ans.
```

Ans. 28620 brick in house.

Paper Hanger's Measurement.

RULE.

Bring the whole into feet and divide by 36, as there are that many square feet in a piece of wall paper 24 feet long and 18 inches wide.

EXAMPLE.

A room 16 feet long, 12 feet wide and 9 feet high. How many pieces of wall paper does it take to cover the same?

```
16 feet long.          12 feet wide.
 9  "  high.            9
 ──                     ──
144  "   one side.     108 feet one side.
  2                      2
 ──                     ──
288 feet two sides.    216 feet two sides.
 ──                    288  "    "    "
Ceiling                 ──
 16 feet long.         504  "  four  "
 12  "  wide.          192  "  Ceiling.
 ──                     ──
192 feet.              36)696(19⅓
                         36
                         ──
                         336
                         324
Ans.  19⅓ pieces         ──
                          12
                         ──=⅓
                          36
```

Painters' and Plasterers' Measurement.

RULE.

Bring the whole into feet and divide by 9.

EXAMPLE.

A room 16 feet long, 12 feet wide and 9 feet high; how many square yards?

16 feet long.	12 feet wide.
9 " high.	9 " high.
144	108
2	2
288 feet two sides.	216 ft two sides.

Ceiling.

12 feet wide.
16 " long.

72
12

192 feet ceiling.
216 } four sides
288

9)696 square feet.

77 3-9=⅓

Ans. 77⅓ square yards.

Number Bushels Lime Kiln Contains.

RULE.

'Multiply the length of base by breadth of same, and the length of top by the breadth of same; add both together and divide by two, to get the average number square feet the stack contains, then multiply that quotient by the height and divide by 1¼ to get the number bushels lime in kiln, as in 1¼ square feet is contained one bushel. 'Eight tons coal burns 1000 bushels stone lime and from 20 to 23 cords wood.

EXAMPLE.

A stack lime 30 feet long at base, and 18 feet broad at same, and 12 feet long at top and 6 feet broad at same and the whole 10 feet high, how many bushels.

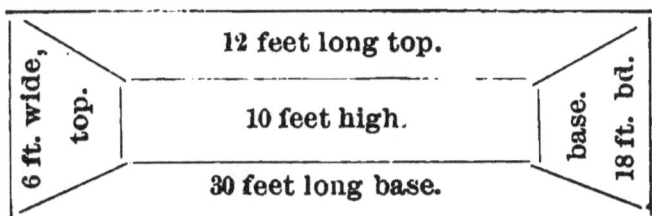

```
6 ft. wide,  |           12 feet long top.          |          18 ft. bd.
      top.   |                                      |   base.
             |           10 feet high.              |
             |           30 feet long base.         |
```

18 feet broad base.   12 feet long top.
 30 "   long   "   6 "   broad "
 ___   ___
 540   72
                            540
                            ___
1¼)   half   612
4-5)3060   ___
        4   306–average sq. ft.
   ____   10 feet high.
  12240   ____
  ____   3060 square feet in
Ans. 2448 bus.   stack.

Number Gallons a Cistern or Well Contains.

RULE.

Square the diameter and multiply that product by 7.854 and that product by the depth; cut off four right hand figures and multiply the remainder by 1728 and divide by 231.

EXAMPLE No. 1.

A cistern 8 feet in diameter and 16 feet deep; how many gallons does it contain?

```
      7.854          8 feet diameter.
        64           8
     ------         ---
     31416           64
     47124
     ------
     502656
          16 feet deep.
     ------
    3015936
    502656
    ------
    804|2496              1728
                          804
                        ------
                          6912
                        13824
                        ------
                  231)1389312(6014 gallons.
                      1386
                      ----
                       331
                       231
                      ----
                      1002
                       924
                      ----
  Ans.  6014 gallons.    78
```

EXAMPLE No. 2

RULE.

If a cistern is square or otherwise multiply the length, breadth and depth together and the product by 1728 and divide by 231.

EXAMPLE.

A cistern 8 feet long, 6 feet broad and 16 feet deep; how many gallons does it contain?

```
8 feet long.            1728
6  "   broad.            768
__                     _____
48                     13824
16 feet deep.          10368
__                     12096
288                   _____
48           231)1327104(5745  galls.
__                    1155
768                   ____
                      1721
                      1617
                      ____
                      1040
                       924
                      ____
                      1164
                      1155
                      ____
                         9
```

Ans. 5745 gallons.

Foundation or Cellar.

RULE.

Bring the whole into feet and divide by 9.

EXAMPLE.

A foundation or cellar 20 feet long, 12 feet broad and 12 feet deep; how many square yards does it contain?

```
        20 feet long.
        12  "   broad.
        ───
        240
        12 feet deep.
        ───
        480
        240
        ───
     9)2880
        ───
Ans.       320 square yards.
```

Carpenters' Rules.

To ascertain the number of feet of weather-boarding a building contains.

RULE.

Multiply the length by the height and the breadth by the same, and to get the number of feet gable end contains, take half the height and multiply by the breadth.

EXAMPLE.

A house 30 feet long, 18 feet high and 24 feet broad and 18 feet high; gable end 8 feet high and 24 feet broad.

```
18 feet high.          24 feet broad.
30  "  long.           18  "   high.
———                    ———
540 one side.          192
  2                    ·24
———                    ———
1080 two sides.        432 one end.
———                      2
24 feet gable end.     ———
 4  "   half.          864 two ends.
—                      1080
96 one gable end.      ———
 2                     1944 feet four sides.
—                       192  "  gable ends.
192 feet gable ends.   ———
                       2136 feet.
Ans.  2136 feet.
```

Carpenters' Rules—Continued.

How to ascertain what pitch to give a roof, &c.

RULE.

To get the pitch of a roof or to get the length of the rafters, take two-thirds of the breadth of the building, and to get the height of the gable end take one-third of the breadth of the building.

EXAMPLE.

If a building is 24 feet long two-thirds of 24 would be 16, then the rafters would be 16 feet long and one-third of 24 would be 8, then the gable ends would be 8 feet high.

Carpenters' Rules—Continued.

To get the number of shingles a building contains.

RULE.

Multiply the length of the building by the length of the rafters, to get the number square feet, then multiply that product by the number shingles it takes to cover a square foot. Joint shingles 6 inches wide it will take 4, and same 4 inches wide it will take 6, and oak lap shingles 3. In covering with boards get the number square feet and add one-third more for lap.

EXAMPLE.

If the rafters in a building are 16 feet long and building 24 feet long and shingles 6 inches wide, how many shingles to cover both sides of house.

```
 16 feet long rafters.
 24  "    "   building.
 ──
 64
 32
 ──
384
  4 shingles to square foot.
────
1536 one side.
   2
────
3072 shingles Ans.
```

Carpenters' Rules—Continued.

To ascertain the length of rafters for a building.

RULE.

For one-third pitch multiply width of building by 60, and one-fourth pitch multiply same by 56, and one-fifth pitch multiply same by 54, and one-sixth pitch multiply same by 53 and in all point off two figures from the right and add to the lengths for projection.

EXAMPLE No. 1.

What would be the length of the rafters if a house is 24 feet wide and you want to give it one-third pitch.

$$\begin{array}{r} 24 \\ 60 \\ \hline 14{\mid}40 \end{array}$$

Ans. Rafters 14 feet.

EXAMPLE No. 2.

Length of rafter, same width and one-fifth pitch.

$$\begin{array}{r} 24 \\ 54 \\ \hline 96 \\ 120 \\ \hline 12{\mid}96 \end{array}$$

Ans. 12 feet   12|96

Carpenters' Rules—Continued.

Number of feet a floor contains.

RULE.

Multiply the length of floor by breadth of same and if more than an inch in thickness add accordingly.

EXAMPLE.

A room 2 feet long and 14 feet wide; how many feet inch boards does it contain, and how many one and a quarter inch.

```
 14 feet wide.
 20  "  long.
 ___
280 feet inch boards.
 70
 ___
350 feet 1¼ inch boards.
```

Size of Nails.

The following table will show at a glance, the length of the various sizes and the number of nails to the pound.

3 penny, 1   inch 557 to the pound.
4   "   1¼   "   353 "   "   "
5   "   1¾   "   232 "   "   "
6   "   2   "   167 "   "   "
7   "   2¼   "   141 "   "   "
8   "   2½   "   101 "   "   "
9   "   2¾   "   85 "   "   "
10   "   3   "   68 "   "   "
12   "   3¼   "   54 "   "   "
20   "   3½   "   34 "   "   "

Fence Nails.

7 penny, 2   inch long 80 to the pound,
8   "   2¼   "   "   70 "   "   "
9   "   2¾   "   "   50 "   "   "
10   "   3   "   "   40 "   "   "

Spikes.

4   inch, 16 to the pound.
4½   "   12 "   "   "
5   "   10 "   "   "
6   "   7 "   "   "
7   "   5 "   "   "

Wheat, Corn &c. Measurement

RULE.

Ascertain the number of pounds of wheat, corn &c., and to get the number of bushels, divide by the standard weight. The standard weight of wheat is 60 pounds; corn and rye 56 pounds; barley 48 pounds and oats 32 pounds.

EXAMPLE.

If 50 bags wheat weigh 6000 pounds gross, and bags 50 pounds, how many bushels wheat do the bags contain?

```
6000 gross.
  50 bags.
  ───
5950 pounds wheat.
```

```
60)5950(99 bushels and 10 pounds.
   540
   ───
    550
    540
    ───
     10
     ──=⅙
     60
```

99 bushels and 10 pounds wheat at $1.50 per bushel.

```
 1.50            If 60 pounds cost $1.50 what
   99 bushels.   will 10 pounds cost.
 ───               60—150—10
1350                  10
1350                  ──
───            60)1500(25 cents.
148.50            120
  25              ───
 ───              300
$148.75 Ans.      300
```

Land Measure.

To measure off an acre of ground, step or measure 69 yards one way and 70 yards the other way, that will give you 4830 square yards, accurate enough for any practical purpose.

EXAMPLE.

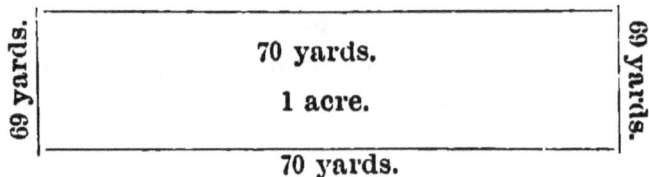

$$\begin{array}{r} 69 \\ 70 \\ \hline 4830 \text{ square yards.} \end{array}$$

If a piece of ground steps or measures off 200 yards one way and 250 yards the other way, then multiply both together and divide by 4840 to ascertain the number of acres.

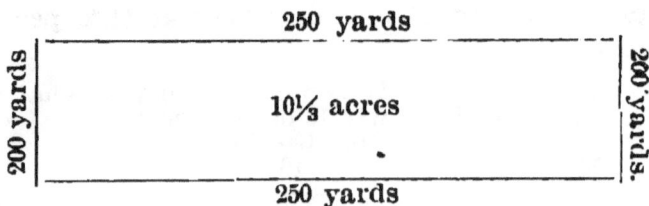

$$\begin{array}{r} 250 \\ 200 \\ \hline \end{array}$$

$$4840)50000(10\tfrac{1}{3} \text{ acres.}$$
$$\underline{4840}$$
$$1600$$
$$\frac{1600}{4849}=\tfrac{1}{3}$$

Land Measure—Continued.

If a field steps off 200 yards at one side and 150 yards at the other side, and 90 yards straight across at both ends, how many acres does the field contain?

RULE.

Add the sides together and take half to get the average length, and multiply by the length of the ends and divide by 4840.

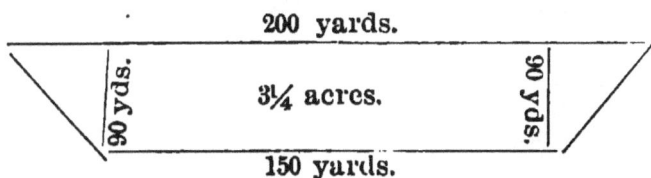

200 yards.

90 yds. | 3¼ acres. | 90 yds.

150 yards.

200 yards.
150 "

half 350
——
175

175 average length sides.
90 feet length ends.
——
4840)15750(3¼ acres.
14520
——
1230
——— =¼
4840

3.

Interesting Tables for Farmers.

Quantity of seed usually sown to an acre.

Timothy	¼	to ½	bushels.
Red Top	½	" 1	"
Red Clover	6	to 10	pounds.
White Clover	5	" 8	"
Lucerne		10	"
Orchard Grass	1	" 1½	bushels.
Blue Grass	1	" 1½	"
Wheat	1½	" 2	"
Barley	1½	" 2	"
Buckwheat	1	" 1½	"
Rye Grass	1	" 1½	"
Carrot	2½	" 3	pounds.
Beet	4	" 6	"
Parsnip	3	" 5	"
Onion	4	" 6	"
Ruta Baga		1	"
Turnip	1	" 1½	"
Beans	1½	" 2	bushels.
Peas	1½	" 2	"
Oats	2½	" 3	"
Rye	1½	" 1½	"
Millet	½	" ¾	"

The quantity of Corn required to plant an acre, 3 grains in the hill.

3 feet	by 2—	11	quarts.
3¾ "	by 3—	6	"
3 "	by 4—	4	"
3 "	by 3—	7	"
3¾ "	by 3½	5	"
4 "	by 4—	3½	"

The Number of Plants per Acre at a given Distance.

1	foot,	53,560	6 feet,	1,210
1¼	feet,	19,360	9 "	537
2	"	10,890	12 "	302
2½	"	6,968	15 "	194
3	"	4,840	18 "	134
4	"	2,722	20 "	109
5	"	1,742	25 "	69

One acre of tobacco set 3 feet by 2½ distant will contain 6,050 plants. Most growers prefer 3¼ feet by 2. This will fill a building 20 by 40 with 12 to 15 feet post.

Interest Calculated at 6 per Cent.

RULE.

Take half the number months and multiply
by the amount you want to find the interest
thereon and cut off two right hand figures. To
find the interest for days, take half the number,
proceed as above and divide by 30.

EXAMPLE.

Interest on $40.25 for 4 years, 9 months and 24
days.

```
    40.25                  4 years and 9 mos.
      28½           half   57 mos.
    _____                 __
    32200                  28½
    8050                   __
    2012
    _____                 $40.25—24 days.
    11.47|12                 12  __
    16                       __   12
    $11.63               30)48300(16|10
                             30
                             __
                             183
                             180
                             ___
                              30
                              30
                              __
    Ans.  $11.63              00
```

*Interest Calculated at 7 per Cent and Up-
wards.*

RULE.

Calculate the interest at 6 per cent. as in for-
mer page ánd add accordingly. If 7 per cent.
add one-sixth more. If 8 per cent. acd one-third
more. If 9 per cent. add ohe-half more and so
on.

EXAMPLE.

Interest on $40.25 for 4 years, 9 months and 24
days at 8 per cent.

```
 $40.25           4 yrs. 9 mos—57
    28½
 ──────            ½)57 months.
 32200             ──────
  8050              28½
  2012             ──────
 ──────            ½)24 days.
 $11.47|12   $40.25  ──
      16         12    12
 ──────      ──────
 $11.63      3|0)4830|0

                16|13
```

That would be $11.63 interest for the time giv-
en at 6 per cent. Now to get the interest at 8
per cent. add one-third more, as 2 is the third of
6, and 6 and 2 are eight.

```
              $11.63  6 per cent.
 Add ⅓         3.88
              ──────
 Ans.         $15.51  8 per cent.
```

Interest Calculated at 6 per Cent.—Another Form.

RULE.

To calculate interest at six per cent. call half the number of months cents, and one-sixth the number of days mills, and their sum will be the interest on one dollar for given time, and after calculating the interest on a sum of money cut off as many figures as you multiply by, and the remainder will be the interest for given time.

EXAMPLE.

Interest on $40.25 for 4 years 9 months and 24 days.

```
                              cts. mills.
4 yrs. 9 mos.—57 mos. ½—     28  .5
   24 days ⅙=                     .4
                              ─────────
                              28.9
```

28 cents and 9 mills on one dollar for given time.

```
        $40.25
         28.9
        ──────
        36225
        32200
        8050
        ──────
        $11.63 225
```
Ans. $11.63

At 7 per cent add one-sixth more. At 8 per cent. add one-third and so on as in preceding page.

To Calculate Interest on a Note When Partial Payments are Made.

RULE.

Calculate the interest up to the time the first payment is made, then add same to principal and deduct payment, then calculate interest on balance to second payment, add to balance, deduct payment and so on.

EXAMPLE.

A note given for $800.00, dated March 1st, 1875, payable 8 months after date, wiih interest from date, with a credit May 1st, 1875, of $300.00. Also one of 400.00 July 15th, 1875. What would be the balance, both princidal and interest, when the note matured.

$800.00 principal.
1 half number of months.

$8.00. 00	2 mos. int. from March 1 to May 1.
800.00	add principal.

$808.00
300.00   paid May 1st, 1875.

508.00   Balance due May 1st, 1875.
1¼   half number months.

508.00
127.00

6.35 00   2½ mons. int.
508.00   add balance.

514.35
400.00   paid July 15th, 1875.

114.35   balance due July 15, 1875.
1¾   half number months.

114.35
85.75

2.00|10
114.35

$116.35   Ans. $116.35.

Percentage.

RULE.

Multiply the profit by 100 and divide by the cost.

EXAMPLE.

An article costing $2.40 and selling for $2.60, the profit then would be 20 cents.

2.40 . cost.   20   profit—100

$$
\begin{array}{r}
100 \\
\overline{} \\
240)2000(8\frac{1}{3} \\
1920 \\
\overline{} \\
80 \\
\overline{} = \frac{1}{3} \\
240
\end{array}
$$

Ans.   8⅓ per cent.

Co-Partnership—Gain.

RULE.

Add together each investment and divide that sum into amount of stock, gain &c., to get percentage of each, then multiply the same by amount invested to get each one's share.

EXAMPLE.

Three partners invest as follows :—A $100.00, B. $75.00, C. $50.00, and at the end of the year the gain amounts to $600.00. What is each one's share of the gain ?

```
A. $100.00
B.   75.00        225.00)600.00(2⅔
C.   50.00               450.00
     ──────
  $225.00                150.00
                         ──────=⅔
                         225.00
$100.00
     2⅔
  ──────
  200.00          Ans.  A's. share $266.66⅔
  66.66⅔                B's.   "     200.00
  ──────               C's.   "     133.33⅓
$266.66⅔ A's.                        ──────
  ──────                             $600.00
   75.00
     2⅔
  ──────
  150.00
   50.00
  ──────
$200.00 B's.

   50.00
     2⅔
  ──────
  100.00
   33.33⅓
  ──────
$133.33⅓ C's.
```

Co-*Partnership—Loss*.

RULE.

Divide investment into amount, stock &c., on hand and deduct that quotient from 100, then multiply that by each partner's investment to ascertain the amount of each one's loss,

EXAMPLE.

A. invests $400.00, B. $250.00 and C. $100.00, and at the end of one year, the stock &c., on hand amounted to $300.00. What did each partner lose?

```
        A.  $400.00
        B.   250.00
        C.   100.00
            ───────
        $750.00—300.00   100
                         100
                        ────
              750,00)3000000(40
                      300000
  100                 ──────
   40                      0
  ──
   60          400.00
                   60
              ────────
              $240.00,00  A's.
  250.00                      100.00
      60                          60
  ─────────                   ─────────
  $150.00|00 B's.             $60.00|00 C's.

        Ans.  A.  Loses  $240.00
              B.    "     150.00
              C.    "      60.00
                          ───────
                          $450.00
```

To Ascertain the Amount Actually Made When Selling a Certain Percentages.

RULE.

Add 100 to the average per cent. and divide into the amount of sales; that will give you the cost of the goods. To find the amount made, deduct the cost from the amount of sales.

EXAMPLE No. 1.

If you sell $40.00 worth of goods and they average you 20 per cent., how much of the $40.00 did you pay for the goods and how much is actually made on the sale?

```
     1.20          40.00          100
                    100
                  ───────
           1.2 0)40000 0        $40.00
                               33.33⅓
Cost of goods,    $33.33⅓       ───────
          Amount actually made,  6.66⅔
                               ───────
```

EXAMPLE No. 2

Sold $90.00 worth of goods and they averaged 16⅔ per cent.

```
        1.16⅔  90.00—100
            3     100
        ───    ────────
        350  )90.0000
                 3

        350)2700000(77.14 2-7 cost of goods.
            2450
            ────            $90.00
            2500            77.14 2-7
            2450            ─────────
            ────            $12.85 5-7
             500      Amount actually
             350      made on sale.
            ────
            1500
            1400
            ────
             100
             ───=2-7
             350
```

To Ascertain How Much is Lost When Selling Goods at Certain Percentages Below Cost.

RULE.

Deduct the amount of percentage you are selling at below cost from 100, and proceed as in preceding rule.

EXAMPLE No. 1.

It you sold $40.00 worth of goods and you lost 20 per cent., how much did the goods cost?

```
100
 20 per cent. lost.
 ---
 80           40.00—100
                  100
            ----------------
          8|0)40000,0
            ----------------
            $50.00  cost of goods.
             40.00  amount of sales.
             ------
            $10.00  amount lost.
             ------
```

EXAMPLE No. 2.

Sold $80.00 worth goods and lost 40 per cent.

```
100
 40
 ---
 60           $80.00—100
                  100
            ----------------
          6,0)80000|0
            ----------------
            $133.33⅔  cost of goods.
             80.00    amount of sale.
             ------
            $53.33⅔   amount lost.
             ------
```

To Ascertain What Per Cent. Stocks Pay When They Are Selling Above Par Value.

RULE.

Calculate the interest a single share is paying at par value and divide the selling price into the same.

EXAMPLE No. 1.

If stocks are selling at $22.50 per share and the par value is $15.00, and said stock pays 12 per cent. on the par value. What per cent does it pay the buyer.

$15.00   par value.
   12   per cent.
———
$1.80.00

22.50)1.80.00(8 per cent. pays.
   1 80 00

Ans. Pays the buyer 8 per cent.

EXAMPLE No. 2.

Stock selling at $18.00 and par value $15.00, and pays 10 per cent. on par value.

$15.00   par value.
   10   per cent.
———
18.00)1.50.00(8⅓
   1 44 00
———
  600
——— = ⅓
1800

Pays the buyer 8⅓ per cent.

Mixed Numbers Multiplied.

RULE.

Reduce to improper fractions and multiply numerators together for new numerators, and denominators together for new denominators.

EXAMPLE. No. 1.

Multiply 18¾ by 18¾.

$$\frac{18\frac{3}{4}}{4}$$

$$\frac{75}{4} \quad \frac{75}{4} \times \frac{75}{4} = \frac{5625}{16} = 351\frac{1}{2}$$

EXAMPLE No 2.

Multiply 20⅓ by 18 1-5

$$\frac{20\frac{1}{3}}{3} \quad \frac{18\ 1\text{-}5}{5}$$

$$\frac{61}{3} \times \frac{91}{5} = \frac{5551}{15} = 370\ 1\text{-}15$$

EXAMPLE TO PROVE.

$$\begin{array}{r} 2033\frac{1}{3} \\ 1820 \\ \hline 40660 \\ 16264 \\ 2033 \\ 606\frac{2}{3} \\ \hline 370.06\ \vert\ 66\frac{2}{3} \end{array}$$

Division of Fractions.

RULE.

Invert the divisor and proceed as in multiplication of fractions.

EXAMPLE.

Divide 4 2-5 into 40¾

$$\frac{4 \; 2\text{-}5}{5} \qquad\qquad \frac{40\frac{3}{4}}{4}$$

$$\frac{22}{5} \qquad\qquad \frac{163}{4}$$

$$\frac{5}{22} \times \frac{163}{4} = \frac{815}{88}$$

$$88)815(9\frac{1}{4}$$
$$792$$

$$\frac{23}{88} = \frac{1}{4}$$

EXAMPLE TO PROVE.

$$440(4075(9\frac{1}{4}$$
$$3960$$

$$\frac{115}{440} = \frac{1}{4}$$

Mechanic's Lien.

A Lien may be filed within six montes after the work is completed or material furnished, providing the party is owner of the ground on which the building is constructed. But if such is owned by another party, then the lien must be filed within sixty days after the building is completed or material furnished, and the party who is owner of such ground must be legally notified that such a lien is standing against such property.

FORM.

GEORGE DEBTOR,

TO SAMUEL CREDITOR:

To a quantity of Lumber, Work or Building Material, (as the case may be) as hereinafter specified, amounting to——dollars, with a credit of —— dollars (if there be such) leaving a balance of —— dollars, furnished at the times hereinafter mentioned, to the said George Debtor and used by him in the building or construction of a certain house (or whatever the building represents,) situated (here describe the situation.) Said building is —— feet high——feet long and —— feet broad, of which the said George Debtor is owner, and which building was finished on or about the —— day of——18 .

Mechanic's Lien—Continued.

Here itemize your account, day and date.

Then make oath to the following:

STATE OF MARYLAND,

County, to wit:

On this —— day of ———— 18 , before the subscriber, a Justice of the Peace of the State of Maryland, in and for the said county, personally appeared —— —— and made oath on the Holy Evangelly of Almighty God that the above account is just and true, and that he has not received any part or parcel of the money charged as due by said account, or any security or satisfaction for the same, to the best of his knowledge and belief.

Sworn before

Then have the same recorded in the County Clerk's Office.

4.

*Square Logs Reduced to Inch Board Meas-
ure.*

RULE.

Multiply the breadth by the breadth and the
product by the length, then divide by 12 and deduct one-fifth.

EXAMPLE.

A log 18 x 18 inches square and 16 feet long,
how many feet inch boards does it contain?

```
            18 inches
            18   "
           ———
           144
            18
           ———
           324
            16 feet long
          ————
          1944
           324
          ————
      12)5184
          ————
Deduct 1-5)  432
             86
            ———
Ans.        846 feet inch boards.
```

In length.	Square Logs Reduced to Inch Board Measure.						
feet	10x 10	11x 11	12x 12	13x 13	14x 14	15x 15	16x 16
10	66	80	96	112	131	150	170
12	80	97	115	135	157	180	205
14	93	113	134	158	·183	210	239
16	107	129	154	180	209	240	273
18	120	145	173	202	235	270	307
20	133	161	192	225	261	300	341
22	146	177	211	247	237	830	375
24	160	194	230	271	314	360	410
26	173	210	250	293	339	390	443
28	186	226	269	315	366	420	478
30	200	242	288	338	392	450	512

Feet in length.

Square Logs Reduced to Inch Board Measure.

	17 x 17	18 x 18	19 x 19	20 x 20	21 x 21	22 x 22	23 x 23	24 x 24	25 x 25
10	192	216	240	266	294	322	352	384	416
12	231	259	289	320	353	387	423	461	500
14	270	303	337	373	411	451	494	538	583
16	308	346	385	426	471	516	564	615	667
18	346	389	433	480	529	581	635	691	750
20	385	432	481	533	588	645	705	768	833
22	423	476	529	586	646	710	775	845	916
24	462	519	578	640	706	775	847	922	1000
26	501	562	626	693	764	839	917	999	1083
28	539	605	674	747	824	903	987	1075	1167
30	578	648	722	800	882	968	1058	1152	1250

Round Logs Reduced to Inch Board Measure.

EXPLANATION.

This rule has been thoroughly tested by the Author at the Steam Saw and Planing Mill of Gruber & Witmer, Williamsport, Md., by getting the mean diameter of logs before they went on the mill, and measuring the actual number of feet the same logs made when they were sawed, and found this rule as accurate, of not more so, than the majority of rules, and which rule will do justice to both buyer and seller.

RULE.

Square the mean diameter and multiply that product by 7.854 ; cut off four right hand figures, then multiply by the length and divide by 12, deduct one-third and the residue will be the amount in feet of inch boards.

Round Logs Reduced to Inch Board Measure.

EXAMPLE.

A log 14 inches mean diameter and 20 feet long, how many feet of inch boards.

```
14 inches              7.854
14    "                 196
--                     -----
56                     47124
14                     70686
--                      7854
196                    ------
                      153|9384
                        20 feet long
                       ----
                    12)3060
                       ----
Deduct        ½) 255 .
                   85
                  --
        Ans.      170 feet inch boards.
```

Length in feet.	*Round Logs Reduced to Inch Board Measure.*						
	Mean Diameter.						
	Inch. 12	Inch. 13	Inch. 14	Inch. 15	Inch. 16	Inch. 17	Inch. 18
10	63	73	85	98	112	126	141
12	75	88	102	117	134	151	170
14	88	103	119	137	156	175	198
16	101	118	136	157	179	200	226
18	113	132	153	176	201	226	254
20	126	147	170	196	223	251	282
22	139	161	188	216	246	276	310
24	151	176	204	235	268	302	339
26	164	191	223	255	290	326	367
28	177	206	238	275	313	352	395
30	190	220	255	293	335	377	423

Length in feet.

Round Logs Reduced to Inch Board Measure.

Mean Diameter.

	Inch. 19	Inch. 20	Inch. 21	Inch. 22	Inch. 23	Inch. 24	Inch. 25
10	158	174	192	211	231	251	272
12	189	210	231	253	277	302	327
14	220	244	269	295	323	351	381
16	251	279	307	338	369	4C1	436
18	284	314	346	380	415	452	490
20	314	346	384	422	461	502	544
22	346	384	423	464	507	552	599
24	378	419	461	507	553	603	653
26	409	453	500	545	600	653	707
28	440	488	538	591	645	703	762
30	472	523	577	633	691	753	817

Round Logs Reduced to Inch Board Measure.

Mean Diameter.

Length in feet.	Inch. 26	Inch. 27	Inch. 28	Inch. 29	Inch. 30	Inch. 31
10	295	318	342	367	393	419
12	353	381	410	440	471	503
14	412	445	478	514	549	586
16	471	508	547	587	627	670
18	530	572	615	660	706	754
20	589	635	684	734	784	838
22	647	699	751	807	863	922
24	708	753	820	880	942	1006
26	766	826	888	953	1021	1090
28	824	889	957	1027	1098	1174
30	883	953	1025	1100	1177	1258

Round Logs Reduced to Inch Board Measure.

Mean Diameter.

Length in feet.	Inch. 32	Inch. 33	Inch. 34	Inch. 35	Inch. 36
10	447	475	504	534	565
12	536	570	605	641	678
14	625	665	706	748	791
16	715	760	807	855	904
18	804	855	908	962	1017
20	893	950	1009	1069	1130
22	982	1045	1110	1176	1243
24	1071	1140	1210	1283	1356
26	1160	1235	1311	1390	1469
28	1249	1330	1412	1497	1582
30	1340	1425	1512	1604	1695

www.ingramcontent.com/pod-product-compliance
Lightning Source LLC
Chambersburg PA
CBHW022036080426
42733CB00007B/847